THE CIA LOCKPICKING MANUAL

THE CIA LOCKPICKING MANUAL

CENTRAL INTELLIGENCE AGENCY

Skyhorse Publishing

Skyhorse Publishing books may be purchased in bulk at special discounts for sales promotion, corporate gifts, fund-raising, or educational purposes. Special editions can also be created to specifications. For details, contact the Special Sales Department, Skyhorse Publishing, 307 West 36th Street, 11th Floor, New York, NY 10018 or info@skyhorsepublishing.com.

Skyhorse® and Skyhorse Publishing® are registered trademarks of Skyhorse Publishing, Inc.®, a Delaware corporation.

www.skyhorsepublishing.com

10 9 8

Library of Congress Cataloging-in-Publication Data is available on file.

ISBN: 978-1-61608-232-1

Printed in China

CONTENTS

INTRODUCTION

There has been much opinion and little fact written on the subject of lock picking. It will be my purpose to clarify the facts about this process and at the same time train you in proper procedure so that before you leave this class today, you will at least have picked one lock. Please note that to become truly proficient you must devote much time and patience in the future.

In this volume we will discuss not only the fundamental theories of lock picking but proper terminology, the importance of tool design (using the right tool for the right job), the effects of tolerances, and finally the techniques most commonly used by locksmiths to successfully pick the vast majority of standard pin and wafer tumbler locks.

CHAPTER ONE

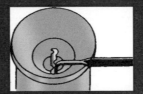

BASIC LOCK PICKING

irst of all, lock picking must be divided into two categories which are commonly confused:

The first category is picking (the act of carefully ma-
nipulating one pin at a time for the expressed purpose of
duplicating the action of the proper cut key in a given cylinder, by something other than the proper cut key itself).

The second category is raking (the less specific act of taking a raking instrument for the expressed purpose of gliding the tool across tumblers of approximately the same depth in a general yet sequential fashion).

Either of these techniques is intended to be a method of convenience for opening locks in emergency type situations. Obviously, the method for making keys referred to as impressioning would be far more desirable since both processes take about the same time and only one yields both an open lock and a working key. However, there are times when picking is the most logical method to use (i.e. when someone is locked out of a house or car and the keys are inside). Both methods are predicated on their efficiency and, should either take an undue amount of time, it is questionable how worthwhile they are when a method such as drilling is so quick and sure, though more expensive.

In order to understand how to compromise a lock there are certain steps which are essential to laying a proper foundation. They are: A thorough working knowledge of the lock mechanism, how it functions, and the ability to recognize these factors so that you are able to overcome them.

CHAPTER TWO

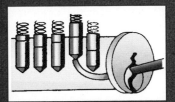

THE CONCEPTS BEHIND
LOCK PICKING

ave you ever thought, "Just what is it that actually allows us to pick a lock?" It is the inability of the manufacturer of any product to machine parts to an almost flawless level of tolerance. Even if they could reasonably approach their goals, the expense alone would be astronomical.

Therefore, we, as locksmiths, are able to pick a lock, so to speak, due to the reality of this situation. To see specifically what is involved, we must look at a typical cylinder.

The tolerance inadequacies to which I refer can be categorized for easy reference. The first is the difference between the plug and the shell. An acceptable amount of difference is approximately **.005** or about **.0025** all around the plug (see Figure **1**).

FIGURE 1.

Space around the plug

The process by which the keyway is "cut" into the plug is called broaching. This process is easily observed when a blank or cut key is inserted in the keyway and "play" is felt due to a significant tolerance differential.

Probably the most significant problem of this sort is the drilling of the chambers. This takes three forms: Plug diameter differential (Fig. **3**), off-center chambers (Fig. **4**), and concentricity (Fig. 5). This is caused by the cost effective but necessarily imperfect process used to manufacture these cylinders, namely gang drilling—a process by which you drill all the chambers at once—and sequential drilling where you drill one chamber after the other. (See illustrations.) In either case, both methods are imperfect because the drill bit itself changes a microscopic amount each time it is used to drill a chamber. It is no surprise then, that after a hundred or a thousand holes the diameter and the centering functions based on its original diameter are no longer accurate. However, in deference to the manufacturer, he could not possibly stay in business and change the bit for each hole of set of holes. We are therefore left with a necessary evil, but one which we can use to great advantage.

FIGURE 2

"Play" felt due to tolerance differential between blank and keyway

FIGURE 3

Diameter differential
between plug and hole
cut into shell

FIGURE 4

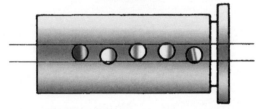

Chambers are slightly off-center due to imperfect drilling procedure

FIGURE 5

Pin clearance

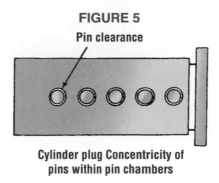

**Cylinder plug Concentricity of
pins within pin chambers**

When turning tension is applied to the core (without the proper key inserted) again tolerance plays a large role in the next operation... Not all of the pins will bind at the same time. Locate those pins, lift them to the correct position (shear line). Follow by doing the same to the next pins to reach the cylinder housing. The only objects which keep the lock from opening are the pins.

CHAPTER THREE

PICKING PROCEDURE

n order to have your best chance to pick a given cylinder, you must not only be aware of the information that
we have provided, but be able to properly utilize it. First,
ascertain whether or not the cylinder can be picked. Does
it operate? Can you manipulate each individual group
of pins within each pin chamber? If you can, then by all
means proceed with the picking and/or raking process. If

not, there is another alternative if you still intend to pick the lock. This problem is more common than you might imagine. Having set your mind upon picking the cylinder, but faced with the problem of "frozen" pins in one or more chambers, the best strategy is to clean and lubricate the lock. This can be accomplished several ways. The preferred methods are shown in figure 6 and figure 7.

FIGURE 6

Spray the keyway to clean and lubricate the lock

FIGURE 7

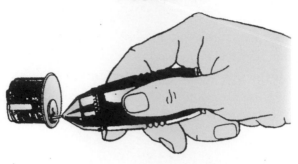

Lubricating graphite gun loosens tight wafers and pins

NOTE: After the application of any solvent or lubricant, impressioning will become difficult, if not impossible.

CHAPTER FOUR

TOOL DESIGN

ool design is a direct result of the function it will be required to perform, and falls into one of three major categories: The hook tool, used when the adjacent bottom pin lengths are significantly different (i.e., 72618). This tool is advantageous for this type of situation, as it allows you to get behind the larger pins in order to properly reach the smaller ones and manipulate them open.

The diamond pick, which is advantageous due to its design in the manipulation of wafer tumblers, which are more fragile and spaced much closer together.

The rake is intended to do just what its name suggests, and is ideal for those situations where all the tumblers are approximately the same size or gradually rise and fall (i.e., 34454, 34565).

FIGURE 8

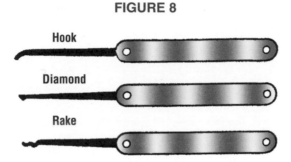

The tools required for raking **are** the rake, the diamond or the ball pick and a tension tool. In this course, I will refer to all raking and picking tools as picks.

FIGURE 9

RAKING TOOLS

FIGURE 9
RAKING TOOLS

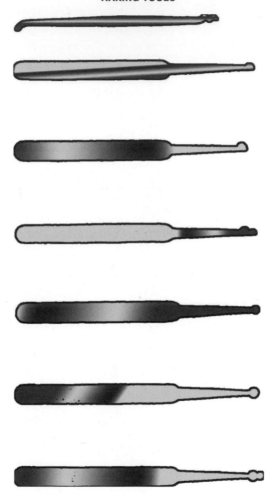

FIGURE 10

RAKING TOOLS

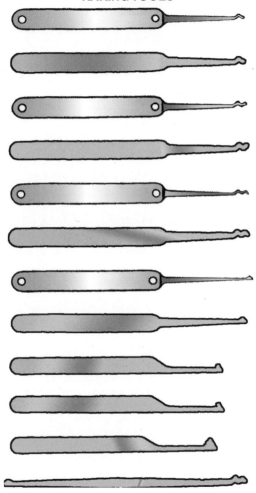

FIGURE 10

PICKING TOOLS

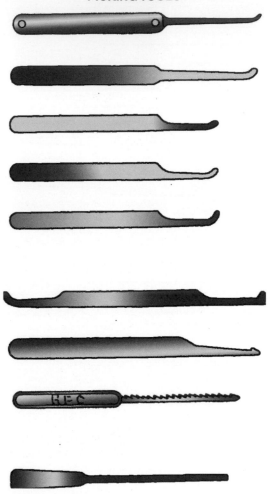

Other individual styles of picks are usually just a modification of one of these groups.

The other tool used in the act of picking is the tension wrench, or more properly, the turning tool. This tool is as or more important than the pick itself but is often overlooked. Too much pressure has defeated more would-be pickers than the wrong type of pick. The main: thing to remember is to use only the lightest amount of pressure necessary to turn the lock. Any more, and you bind the pins so tightly that you make them work against you instead of for you.

Turning tools come in basically 6 groups: light, medium, and heavy duty material and narrow, medium, and wide widths to suit any type of lockpicking situation.

Before you use your tension tool, try raking with the pick a few times. While inserting the pick all the way in the keyway with the tip in contact with the pins, remove the pick with a quick motion keeping an upward pressure on the pins. Repeat this operation again, in slowly and out with a slight snap. Now you are ready to use the tension tool. There are many tension tools to choose from. To start with, I suggest you choose a tool of medium weight and length.

See diagram of other tension tools on the following page

FIGURE 11

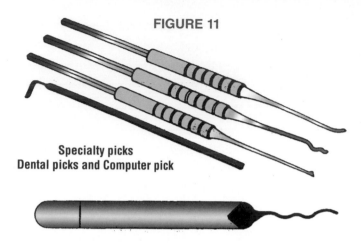

Specialty picks
Dental picks and Computer pick

FIGURE 12

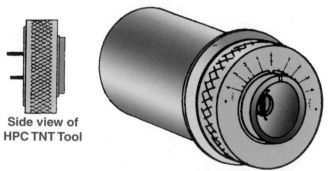

Side view of
HPC TNT Tool

HPC Torque N' Tension Tool in position
over lock cylinder face

Figure 13

TENSION TOOLS

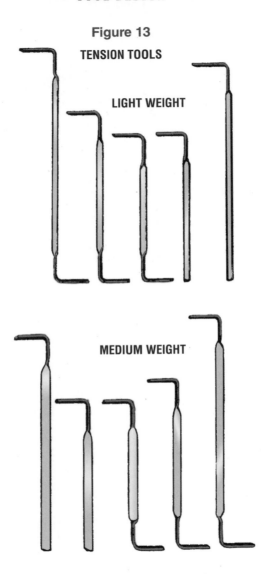

LIGHT WEIGHT

MEDIUM WEIGHT

Figure 13

TENSION TOOLS

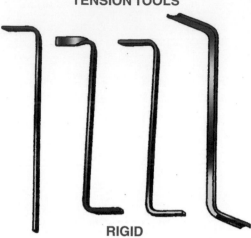

RIGID

DOUBLE SIDED

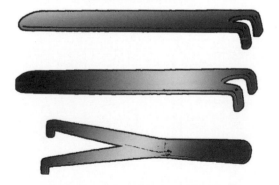

NOTES

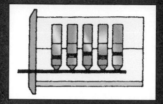

CHAPTER FIVE

MOST COMMON
PICKING TECHNIQUES

The most common techniques for lock picking are raking (where a rake type tool is gently, or in some cases vigorously, pulled along all the tumblers in a rather general way), rather than targeting for specific individual pins as in the case of No. 2, picking each individual chamber. Third is a technique where you would combine the first two. That is, you rake and then specifically target for those pins you may have missed during the initial raking attempts.

Of course, no discussion would be complete without at least mentioning the pick gun. This is a tool that works on the principle of percussion much like cylinder rapping. It is really an effective method once you have mastered the timing necessary to make it work. It consists of the following procedure: Put the tip of the pick gun into the cylinder keyway to be picked. Then, making sure that the pick will strike the pins at a right angle, pull the trigger. At virtually the same moment that the bottom pins are hit, the percussion causes all the top pins to fly straight up towards the

FIGURE 14

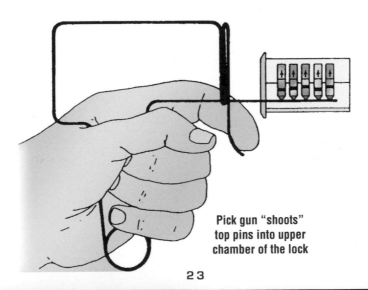

Pick gun "shoots"
top pins into upper
chamber of the lock

top of the pin chamber for an instant, creating an enormous gap. It is in that instant that you must turn the plug with your turning tool, opening the lock. It is this ricochet effect that makes this unique tool so valuable in situations involving specialty pins and cylinders.

NOTE: Specialty items will be discussed in the next book, entitled *Advanced Lock Picking.*

CHAPTER SIX

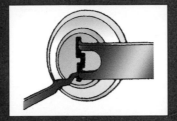

THE LOCKPICKING
PROCESS

uring the lockpicking process, the barest amount of
turning pressure is exerted while you "feel" the condition of the pins in the chambers.

Note: While the ideal condition is matching top and
bottom pins in each chamber to maintain the same pressure in each chamber to insure the best possible cylinder
operation, only purists would say that this is essential, as
evidenced by the fact that almost all of the major lock

manufacturers have gone to a universal top pin. However, such specialty items as mushroom, spool and serrated pins still have to be taken into account and treated as special cases (see Figure 15).

There are three conditions in which you can find the pins in any given chamber once you have ascertained that the lock is operating properly and is therefore pickable. The pins, due to the problems with tolerance differentials acquired unavoidably during the manufacturing process,

FIGURE 15

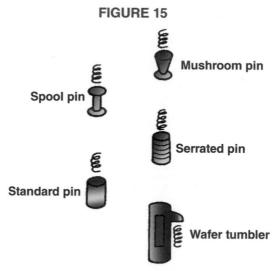

Spool pin

Mushroom pin

Serrated pin

Standard pin

Wafer tumbler

Specialty pins along with standard pin and wafer tumbler

will pick only one at a time no matter how short that span of time may be.

Upon doing your initial raking, the first condition is that the pin is in the unpicked position (see Figure 16).

The second possibility is that the pins in the chamber are merely bound (see Figure 17).

The final possibility is that the pins in the chamber are under pressure, but not bound up (see Figure 18).

Simply continue the process of analyzing the condition of each chamber until they are all picked and the lock is opened.

FIGURE 16

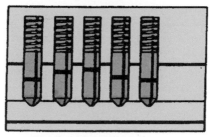

Pins are in the unpicked position

NOTE: The pins wil not necessarily pick in their regular order. By this, I mean that pin number 1 will not necessarily pick first. Perhaps, pin No. 1 will pick fifth and pin No. 3 will pick first, and so on.

Raking is the most common method used today. It is the fastest to use and the quickest to learn. The raking method will work in opening most cylinders where there is not a sudden change in pin sizes, such as a combination of 7-2-6-1-8, where there is one long pin, one short pin, one long pin, and so on.

FIGURE 17

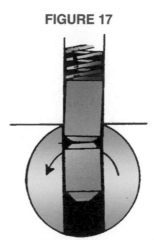

Pins in the chamber are bound

FIGURE 18

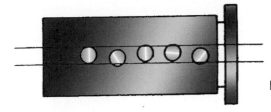

Pins are not bound up but are under pressure due to off-center chambers

FIGURE 19

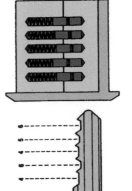

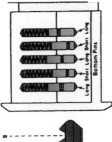

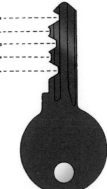

Best cylinder for raking—Similar pin sizes

Worst cylinder for raking—Large variances in pin sizes

The pick you choose for raking should be able to move in and out freely in the upper half of the keyway so it will come into contact with all pins.

FIGURE 20

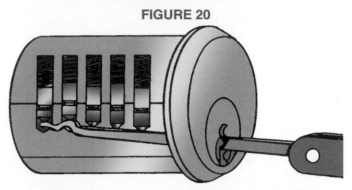

Pick is able to come into contact with all pins

FIGURE 21

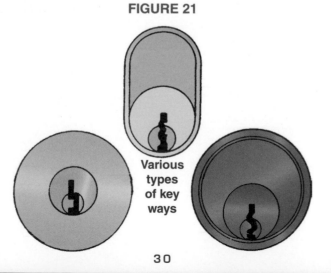

Various types of key ways

The tension tool and its use are the whole trick to raking or picking. Insert tension tool into the bottom of the keyway.

FIGURE 22

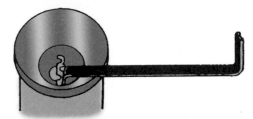

Grooved ends, rigid tension tool

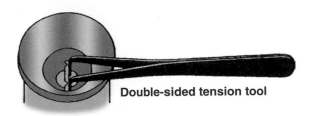

Double-sided tension tool

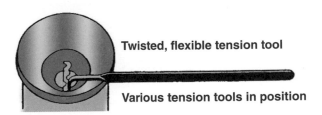

Twisted, flexible tension tool

Various tension tools in position

Then apply very light tension in the direction to unlock the lock. I stress the point: do not use too much tension. You must develop a light touch with the hand that applies the tension. If tension is too heavy, the top pins will bind below the shear-line and will not allow the breaking-point to meet the shearline.

Now, with light tension applied, go through the raking operation, in slowly and out with a snap with upward pressure on the pins with the tip only.

FIGURE 23

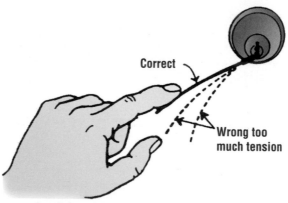

Applying correct tension to the cylinder

FIGURE 24

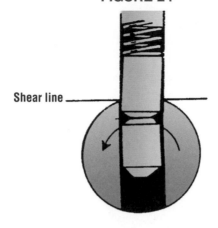

Shear line

Top pins bind below the shear line when tension is too heavy

FIGURE 25

Light tension applied

Repeat this operation three or four times. If the plug does not turn and open the lock, release the tension on the plug—but, before releasing tension, put your ear close to the cylinder and listen for the sound of the pins clicking back into the down position. Release tension slowly so you can hear all the pins. If there is no sound, you were applying too little or too much tension, not allowing the breaking-point to bind at the shear-line.

FIGURE 26

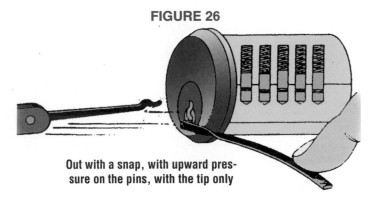

Out with a snap, with upward pressure on the pins, with the tip only

Repeat the raking operation varying the tension, somewhat lighter or heavier than on the first try. With practice, you will gain the right touch in applying tension, and you will find that you can open most cylinders in a few rakings. I suggest you set up a cylinder with only a two pin combination to start with for practice. You should have the cylinder on a large mount, on a door, or held firmly in a vise.

FIGURE 27

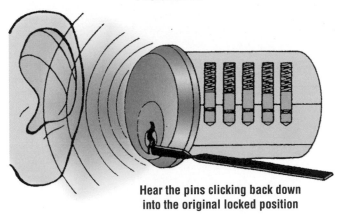

Hear the pins clicking back down into the original locked position

Do not try holding the cylinder in your hand while raking it. After you have conquered the two pin combination, go on to a three pin and so on, until you can rake a six or seven pin cylinder. I have found that in some cylinders, where I have tried raking the regular tension and had no luck in opening them, I would then use a slight pulsating tension, but again—**not too heavy.**

When using the pulsating tension, go from very light to a medium amount of torque, but at all times, when pulsating to the very light, do not lose tension on the pins completely. You will find, in raking cylinders, some will open very easily, regardless of the pin combination. This is

FIGURE 28

Lock cylinder placed in vise

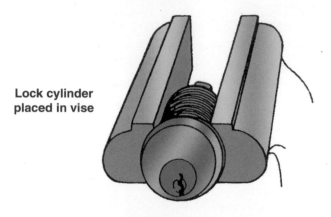

FIGURE 29

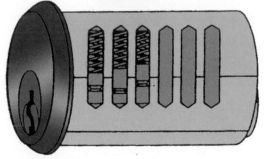

Example of three pin combination

due to the poor construction of some cylinders. As a rule, the lower the price of the cylinder, the easier it picks. The low-priced cylinder is manufactured with greater clearances on all parts so that the cost of assembly will be kept low. The following characteristics are commonly found in low-priced cylinders; too much chamfer on the top of the bottom pin; die-cast plug and body with poor hole alignment; and, over-sized pin holes; too much clearance between plug and body.

FIGURE 30

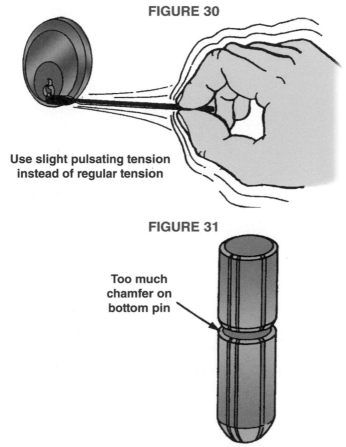

Use slight pulsating tension instead of regular tension

FIGURE 31

Too much chamfer on bottom pin

This is an aid for the manufacturer in the assembly of the cylinder, but it is also an aid for the locksmith who must pick the cylinder. Higher-priced cylinders are manufactured with much less clearance. They are usually constructed from brass bar stock, both body and plug. The pin holes are drilled and reamed for a close fit with the pins and when the plug and body are drilled while together the hole align-

FIGURE 32

Pin clearance

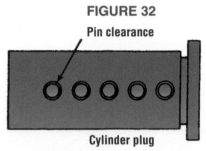

Cylinder plug

Oversized pin holes leave ample pin clearance

FIGURE 33

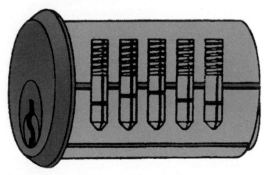

**Clearance be-
tween plug and
body**

ment is excellent. But, in spite of the fine construction, you still can pick or rake it open. It might take a little longer—it might take some adjustment in the tension—it may have to be picked instead of being raked, but you can do it. No matter how minute the clearances are, there are clearances, or the parts would not go together and this is what makes picking and raking possible. At times you will come across a cylinder that you can not pick or rake in a reasonable length of time. Even the expert runs into these same problems. Do not become discouraged. Most locks can be picked or raked in a short time. Do not waste hours working on an extremely difficult cylinder. You will soon be able to determine just how much time to spend on picking or raking a cylinder before resorting to other methods such as drilling.

I have not gone into the handling of the tools. You will probably develop your own personal grip, but for my suggestions see Figure 34.

Try any one of the grips shown in the illustrations. The most important thing to remember at this point, is that the tools must be comfortable in your hands. I suggested at the start that you use a medium weight and length tension tool, but after you have been raking for awhile, you may prefer to try your skill with a light weight or rigid tension tool. You will soon find which tool is best for you.

When raking a lock which has a spring-loaded plug such as most padlocks see Figure 35.

FIGURE 34

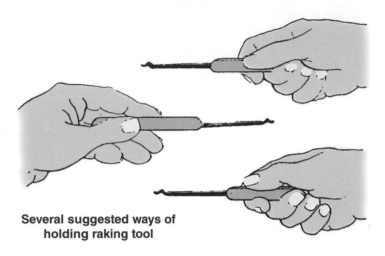

**Several suggested ways of
holding raking tool**

As you apply tension to the plug, you will be working against the direct pressure of **a** spring which is used in the locking of the shackle and returning of the plug to the locked position. This lock will require more tension. Use one of the rigid tools (see Figure 13).

Also if you push inward on the shackle of the padlock, it will relieve some of the spring pressure on the plug (see Figure 36).

FIGURE 35

Cut-away view of spring loaded plug

FIGURE 36

Push downward on shackle to relieve spring pressure on plug

If you discover that you have raked the plug in the wrong direction and the lock will not open, this is no problem. If the lock was very easy to pick, just apply tension in the other direction and re-rake it open. Now, if you have raked the lock in the wrong direction and opened it with difficulty, hold it right there. Do not lock it and re-rake it. There is a tool just for this purpose (see Figure 37)

FIGURE 37

Plug spinners or flip-its

Turning to the left   **Turning to the right**

This pair of coiled springs with handles, one coiled to the left and one coiled to the right, will be referred to as Flip-Its. The first procedure is to determine if the left or right hand Flip-It is required. This is done by facing the lock and visualizing the handle pointing towards you and the flange inserted into the top of the keyway. If the plug is turned to the right, your handle will be to the right of the cylinder—and if the plug is turned to the left, your handle will be to the left of the cylinder (see Figure 38).

Now, say that your plug is turned to the left and this is in the wrong direction for opening the lock. You may

FIGURE 38

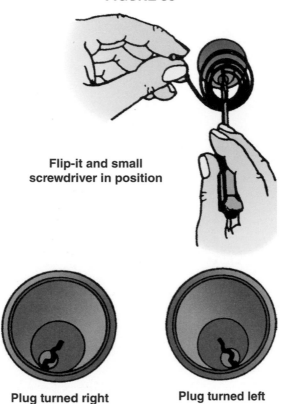

Flip-it and small
screwdriver in position

Plug turned right   Plug turned left

have to get the plug to go to the right. Carefully remove the pick and tension tool. Next insert a small screwdriver into the lower portion of the keyway on the raked lock. Keep the tension on the plug with the screwdriver. Do not allow the plug to slip into the original locked position. You already have chosen the proper flip-it. Now place the centered flange end of the coil into the upper section of the keyway. Grasping the small handle of the coil, strongly wind the flip-it toward the direction into which the plug is to be turned. This will be to the right. At the same time, remember to keep the plug in position firmly with the screwdriver. With a quick yank, pull back the screwdriver. The tension **of** the flip-it will snap the plug over to the opposite direction quickly enough to prevent the pins from falling back into their locked positions. After practicing this procedure, you will find the tool is quite easy to use. Occasionally, you will come across a cylinder that rakes easier in one direction than the other and if you have to rake it into the unlocking direction, you will find this tool quite handy.

Now, for the method of picking. When picking a cylinder, you will be lifting one pin at a time, for this we would use a hook-type pick (see Figure 8).

You apply tension in the same manner as you did when raking. The insert the pick all the way into the keyway and raise up the last pin until the breaking point will bind at the shear-line (see Figure 39).

FIGURE 39

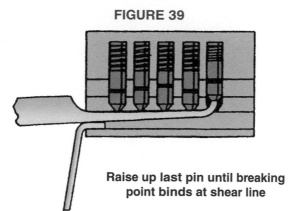

**Raise up last pin until breaking
point binds at shear line**

Then proceed to the next pin until you work your way out of the keyway. Be sure to keep tension on the plug during the entire process. After each pin is picked, you will feel the bottom pin become free of the downward spring pressure. But, don't be fooled just because the bottom pin is free. This doesn't mean it's picked. You may have been applying too much tension to the plug and caused the top pin to bind below the shear-line (see Figure 40).

When it is properly picked, you should feel a very slight give in the turning of the plug (see Figure 41).

That slight turning of the plug will become greater with each pin you pick until it turns fully when you have picked all the pins. Always remember, the tension must not be too heavy. The whole secret of raking or picking is in the

FIGURE 40

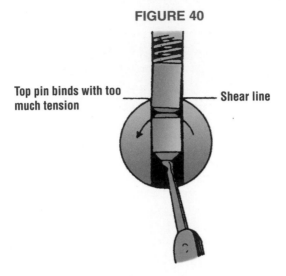

Top pin binds with too much tension

Shear line

tension. You must develop that touch with your tension hand. To practice picking, do so as I suggested in the raking method. Set up a cylinder with two pins, pick it a few times, then set it up with three pins, and so on until you are able to pick a six or even a seven-pin cylinder.

All the previous directions have been for pin-tumbler cylinders. When dealing with the disc-tumbler, see Figure 42.

I find that the raking method is all that is required for opening these locks. The raking is performed in the same manner as that of raking a pin-tumbler cylinder. You will find they rake open quite easily. The tension is also used in

FIGURE 41

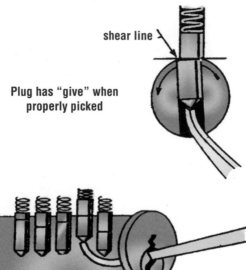

shear line

Plug has "give" when properly picked

Correct tension holds pins at shear line when picking

 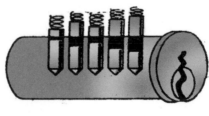

FIGURE 42

**Single sided disc
tumbler cylinder**

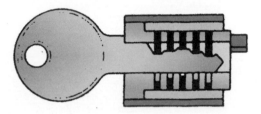

the same manner as that of the pin-tumbler cylinders. The
only variation would be in the double-sided disc or wafer
cylinder (see Figure 43).

The double-sided cylinder usually requires a different
tension tool (see Figure 13).

Double-sided locks can be raked in two different ways.
Number 1—with the use of a standard raking tool—but I
suggest the single or double ball pick which I have found to
work very well. Apply tension in the same manner as you
did with all other picking and raking. However, if you are
using a double-sided tension tool, it will fit in the top and
bottom of the key way (see Figure 44).

FIGURE 43

Double sided disc tumbler cylinder

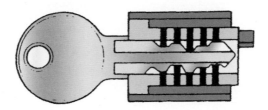

FIGURE 44

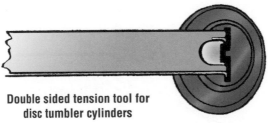

**Double sided tension tool for
disc tumbler cylinders**

After you have applied tension, you begin to rake the upper discs or wafers as you would have in the single-sided lock. When you feel a slight give in the tension of the plug, you switch your raking to the bottom—**but do not let up on the tension.** You rake the lower discs or wafers in the same way as you did the upper ones, but use a downward pressure when pulling the rake out (see Figure 45).

It would be like raking a pin tumbler cylinder that was installed upside-down. Now for the second method of raking double-sided locks. Use double-sided picks (see Figure 46).

You will find these tools very effective in opening most disc-tumbler double-sided locks. With these tools, no tension tool is required if there is no spring tension on the plug. As a general rule, spring tension will be found only in padlocks or shunt switches (see Figure 47).

FIGURE 45

Remove rake with a downward pressure

FIGURE 46

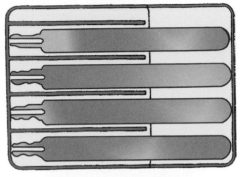

Double sided picking tool set

FIGURE 47

Disc tumbler shunt switch

Disc tumbler padlock

Insert tip of picking tool all the way into the keyway. Rock tool rather rapidly up and down while pushing slowly and gently inward. If pick binds part way in, remove tool and reverse to pick prongs on other side and try again. A pick whose bittings are incompatible with the key-way alignment will not go all the way in. Do not force. When tool is inserted all the way picking action is begun by a moderately rapid up and down rocking motion together with a twisting toward the unlocking direction (see Figure 48).

This up and down rocking should not exceed a one and one-half inch arc at the handle end (see Figure 49).

Combine the rocking and the light twisting motion with a slow in and out raking. Forcing the tool would distort the prongs and may lead to breakage. Each of the

FIGURE 48

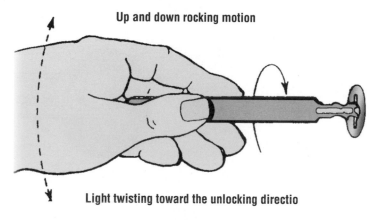

Up and down rocking motion

Light twisting toward the unlocking directio

FIGURE 49

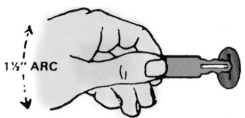

Up and down rocking of tool not to exceed a 1 ½ inch arc

four tools should be tried out using this action. Explore all eight possibilities until a lock opening is made. After removal of the pick, make sure the fork ends are straight. Rebend to a parallel position if it becomes necessary. Most double-sided cylinders unlock to the right, clockwise. When picking the double-sided padlocks or shunt switches a heavier tension is required. We suggest the use of the VV-6 (see Figure 13).

In this case, the tension tool is usually required for these types of locks. Also, the pressing down on the shackle at the same time while raking will help relieve some of the turning tension. You will find both methods work very well after a little practice.

By following these instructions, you should soon become proficient in the raking and picking of pin-tumblers, single and double-sided disc and wafer locks.

NOTES

CONCLUSIONS

L ock picking can be an effective method for opening locks, only if certain conditions present themselves (i.e. the pins must be free and the cylinder, in general, operational). However, we must not let ego get in the way when we determine the best method for a given situation. After all, there is nothing wrong with drilling a lock to get it open, if that is the most time efficient and cost effective method for your customer. Remember, labor is more expensive (in most cases) than product. It is not logical to stand outside picking a lock for one hour, even if you finally do open it, when you can drill and replace most common cylinders in mere moments.

Picking is a skill that will only improve with practice, experience, **and** dedication, but the rewards in a large variety **of** situations will be incalculable in terms **of** time **and money.**